Addressing Changes in Regional Groundwater Resources: Lessons from the High Plains Aquifer

Report of the AGI Critical Issues Forum
October 27–28, 2016, Golden, Colorado

Written by Timothy Oleson, Ph.D.

Contents

- 2   Critical Issues Forum Planning Committee
- 2   Forum Agenda, October 27 and 28, 2016
- 3   Foreword
- 5   Addressing Changes in Regional Groundwater Resources: Lessons from the High Plains Aquifer
- 24   Speaker Biographies
- 29   About the American Geosciences Institute (AGI)
- 30   Selected AGI Projects
- 31   Critical Issues Program

© 2017 American Geosciences Institute
isbn: 978-1974166640
American Geosciences Institute
4220 King Street
Alexandria, VA 22302-1507 U.S.A.
Phone: +1 (703) 379-2480
Fax: +1 (703) 379-7563
agi@americangeosciences.org
www.americangeosciences.org
For more information on Critical Issues Forums, go to
www.americangeosciences.org/policy/ci-forums.

Design by Brenna Tobler, AGI
Cover background image © Sergey Nivens/Shutterstock.com, pipeline image ©Cecilia Lim HM/Shutterstock.com

Critical Issues Forum Planning Committee

Jean Bahr, President, American Geosciences Institute
Rex Buchanan, Director Emeritus, Kansas Geological Survey
P. Patrick Leahy, Executive Director, American Geosciences Institute Foundation
Susan Stover, Outreach Manager, Kansas Geological Survey
David Wunsch, Director/State Geologist, Delaware Geological Survey

Forum Agenda, October 27 and 28, 2016

October 27
Welcome and Introductions
- Paul Johnson, President, Colorado School of Mines
- Allyson Anderson Book, Executive Director, AGI
- Jean Bahr, President, AGI

Overview of Western and High Plains Groundwater Issues and Discussion

Invited Keynote:
Regional Aquifer Changes in the West
- Sharon B. Megdal, Director, University of Arizona Water Resources Research Center

What is the Science Telling us about Regional Aquifers in the West?
- William M. Alley, Director of Science and Technology, National Ground Water Association

Policy and Regulatory Overview of Western Aquifers
- James Eklund, Director, Colorado Water Conservation Board

Economic and Social Importance of Western Aquifers
- Nick Brozovic, Director of Policy, Water for Food Global Institute, University of Nebraska

The High Plains Aquifer — Panel Talks and Discussions

Perspectives from Kansas and Nebraska
- Jim Butler, Senior Scientist, Kansas Geological Survey, University of Kansas
- Ann Bleed, Former Director, Nebraska Department of Natural Resources
- Rex Buchanan, Director Emeritus, Kansas Geological Survey, University of Kansas (moderator)

Perspectives from Texas and Oklahoma
- Steven D. Walthour, General Manager, North Plains Groundwater Conservation District, Texas
- Kyle E. Murray, Hydrogeologist, Oklahoma Geological Survey, University of Oklahoma
- David Wunsch, Director, Delaware Geological Survey, University of Delaware (moderator)

Forum Overview and Moderated Discussion
- John E. McCray, Professor and Head, Civil and Environmental Engineering, Colorado School of Mines
- Wendy Harrison, Professor, Geology and Geological Engineering, Colorado School of Mines (moderator)

Cocktail Reception, Colorado School of Mines Geology Museum

Dinner and After-Dinner Keynote
- Merri Lisa Trigilio, Director and Producer — Written on Water

October 28
Thinking Beyond the High Plains Aquifer

Water and Negotiation in the West
- Susan Stover, Outreach Manager, Kansas Geological Survey, University of Kansas

Invited keynote: Groundwater Policy in the Face of Climate Change
- Jason Gurdak, Associate Professor, San Francisco State University

Lessons, Reports, and Revelations: What Have We Learned?
- Elizabeth Eide, Director, Board on Earth Sciences and Resources, The National Academies of Sciences, Engineering, and Medicine (moderator)
- Wendy J. Harrison, Professor, Geology and Geological Engineering, Colorado School of Mines (moderator)

Concluding Remarks
- Jean Bahr, President, AGI
- Michael Walls, Director, Payne Institute for Earth Resources, Colorado School of Mines

Foreword

The American Geosciences Institute (AGI) represents and serves the geoscience community by providing collaborative leadership and information to connect Earth, science, and people. We created the Critical Issues Forum series as a platform to reach a broader audience of decision makers, including those at the regional, state, and local levels, and to improve public understanding and perception of the geosciences.

I am pleased to present this report summarizing the stimulating presentations and discussions from the second AGI Critical Issues Forum, Addressing Regional Groundwater Resources: Lessons from the High Plains Aquifer. Much has been written about the High Plains Aquifer, due to its critical importance as the major source of groundwater for irrigation in the High Plains region of the United States. This aquifer spans eight states and supports the people and livelihood of region, while also maintaining an agricultural base that is responsible for nearly $35 billion in crops annually. The two-day meeting facilitated lively discussion on common groundwater challenges, resource management approaches, and communication strategies in the High Plains Aquifer region. The forum presentations and dialogue focused on two major questions:

- How have experts and stakeholders in High Plains Aquifer (HPA) states addressed depletion of regional groundwater resources?
- Are there lessons learned or best practices from the HPA and/or other aquifers?

On behalf of AGI and the broader geoscientific community, I extend my sincere thanks to all who participated in the Forum and look forward to hosting other vital conversations highlighting issues that are critical to geoscience and society.

Sincere regards,

Allyson Anderson Book
Executive Director, American Geosciences Institute

AGI thanks the following organizations for their support of the Critical Issues Forum.

The aquifer supplies water for about a quarter of U.S. agricultural production, more than 40 percent of U.S. feedlot beef cattle, and drinking water supplies for 82 percent of the people who live within its boundaries.

Crop circles in Finney County, SW Kansas, taken June 24, 2001. Corn, wheat, and sorghum, plus fallow fields.
Credit: NASA/GSFC/METI/ERSDAC/JAROS, and U.S./Japan ASTER Science Team

Addressing Changes in Regional Groundwater Resources:

Lessons from the High Plains Aquifer

Look out the window of an airplane while in flight over the U.S. High Plains and odds are good — particularly during the growing season — that you'll see swaths of green-hued squares and circles standing out amid otherwise dusty brown landscapes. On the ground, these geometric patchworks are clustered fields of farmland and pasture that both provide a living for many of the people who call these regions home and feed much of the country. These verdant patches are made possible mainly by the presence of groundwater, the lifeblood of irrigation systems in the High Plains region.

Widespread use of groundwater for irrigation in the United States emerged in the early- and mid-20th century, with withdrawals growing for decades subsequent as more — and higher capacity — wells were drilled. Access to abundant groundwater allowed farmers to grow more food on more land and to better withstand crop-withering droughts. The ensuing agricultural boom fed a growing U.S. population and fueled increasing national health, prosperity and food security. Today, roughly 11 percent of U.S. cropland is located in the High Plains Aquifer (HPA) region [Figure 1], and the aquifer supplies water for about a quarter of U.S. agricultural production, more than 40 percent of U.S. feedlot beef cattle, and drinking water supplies for 82 percent of the people who live within its boundaries.

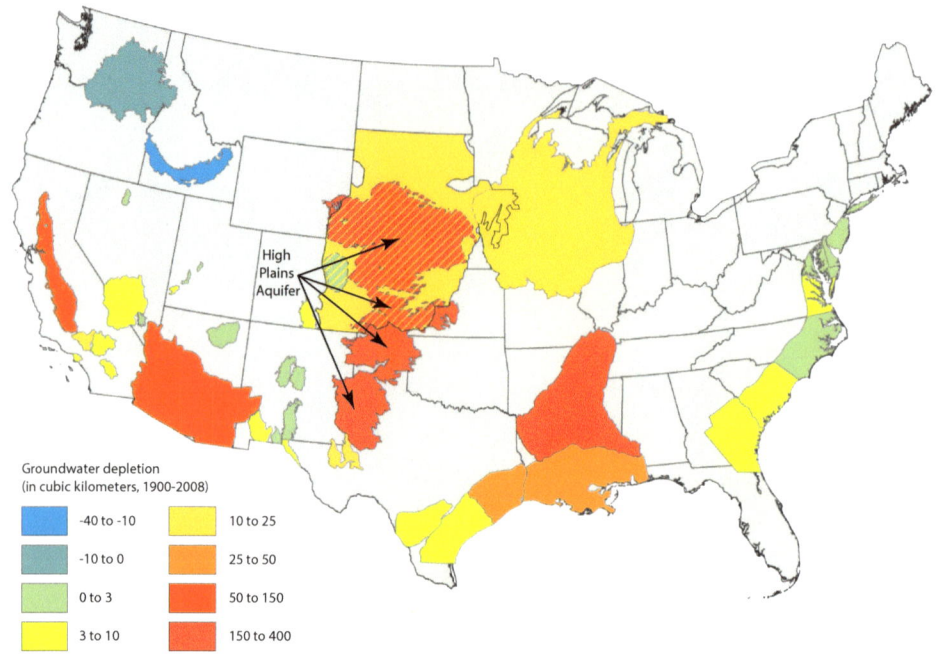

Figure 1: Map of total groundwater depletions (in cubic km) for major aquifers in the contiguous U.S. from 1900-2008. Red 150-400; dark orange 50-150; light orange 25-50; dark yellow 10-25; light yellow 3-10; green 0-3; blues indicate net recharge.
Credit: U.S. Geological Survey Scientific Investigations Report 2013-5079

But along with the prosperity driven by groundwater have come significant concerns about undesired impacts arising from our reliance on the HPA and other aquifers. In particular, due to heavy use and slow recharge of the aquifers, groundwater levels have declined dramatically in many areas [Figure 1], forcing shifts in agricultural practices, jeopardizing the livelihoods of individuals — and whole towns in some instances — and causing collateral damage to the environment. Concerns over groundwater depletion are not limited to the U.S. — major aquifers in China and India have experienced high levels of depletion, for example. Neither are these concerns new. Domestically, however, recent severe droughts in California and in parts of the High Plains, combined with outlooks based on groundwater monitoring data, have brought renewed attention to the fate of the country's most prominent groundwater supplies.

⌜ Along with the prosperity driven by groundwater have come significant concerns about undesired impacts arising from our reliance on the HPA and other aquifers. ⌟

The American Geosciences Institute (AGI) — with generous support from AGI's Center for Geoscience and Society, the Payne Institute for Earth Resources at the Colorado School of Mines, and AGI member societies, including the Geological Society of America, the American Institute of

> **Aquifer:** An underground body of porous materials, such as sand, gravel, or fractured rock, filled with water and capable of supplying useful quantities of water to a well or spring.
> — From https://pubs.usgs.gov/ha/ha747/pdf/definition.pdf.

Professional Geologists, the Association of American State Geologists, the International Association of Hydrogeologists – U.S. National Chapter, the National Association of State Boards of Geology (ASBOG), and the National Ground Water Association — recently convened an open meeting of expert speakers and interested individuals from academia, consulting, professional societies, and local, state and federal agencies to discuss the use, monitoring, and management of groundwater in the United States. The assembled group at this second-ever Critical Issues Forum focused on experiences from the High Plains Aquifer (HPA) region.

The HPA was chosen as the forum emphasis not only because it features prominently in U.S. agriculture and faces significant current and future challenges, but also because — as it extends beneath multiple states — there are a variety of groundwater management practices in use across the High Plains that offer ample opportunities for comparison, information sharing, and learning.

The aim of the forum was to foster open and honest conversation about lessons learned in the region. This report provides an overview of the key lessons and ideas that emerged during the forum, and outlines approaches identified as being potentially beneficial in helping states and municipalities fulfill their own designated groundwater management goals.

The High Plains Aquifer

The High Plains Aquifer stretches across roughly 175,000 square miles (454,000 square kilometers), making it the largest aquifer system in the U.S. [Figure 1], and underlies parts of eight states. The aquifer system comprises buried layers of sand, silt, clay and gravel that were deposited as alluvial sediments, dune sands and valley fill east of the Rocky Mountains starting about 30 million years ago. By far the single largest unit in the HPA is the unconsolidated Ogallala Formation, deposited between about 18 million and 4 million years ago by a shifting network of rivers and streams that carried sediments from the mountains.

The depth to the top of the aquifer differs quite a bit depending on location, ranging from just below the surface to about 400 feet (122 meters), as does its thickness, which ranges from less than 50 to about 1,200 feet (< 15 to 365 meters). These varying characteristics and the diverse geology result in an aquifer with a complex three-dimensional shape — hardly the uniform underground lake that's often, incorrectly, envisioned — which means that water doesn't necessarily flow easily between all parts of the aquifer, and that access to groundwater resources is limited in many places. These factors also affect the cost and effort involved in obtaining groundwater for irrigation and drinking-water supplies, as well as how long local resources will be

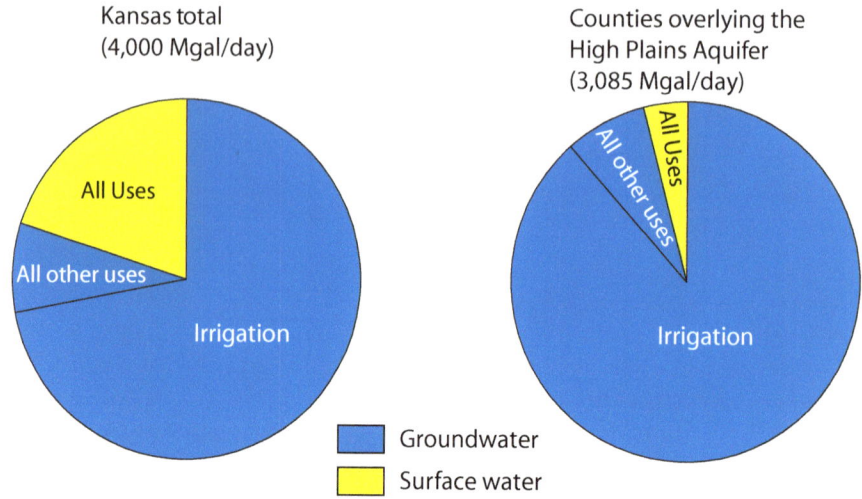

Figure 2: Water Use in Kansas (2010). Groundwater provides 80% of all the water used in the state, and 96% of all water used in counties overlying the High Plains Aquifer. The vast majority of this water is used for irrigation.
Credit: AGI/Ben Mandler. Data source: U.S. Geological Survey

available, particularly in areas where large amounts of groundwater are removed from the aquifer.

Recharge of the aquifer from rainfall, the main counterbalance to withdrawals and discharges, also varies but is low throughout most of the HPA region. Precipitation is actually fairly uniform from north to south across the High Plains, increasing moderately from west to east. But recharge rates differ considerably, from 100 to 200 millimeters per year beneath Nebraska's Sand Hills in the northern High Plains to 10 millimeters or less per year across most of the central and southern High Plains regions. This is due in part to the greater depth to the top of the aquifer, meaning water percolating down from the surface has a longer distance to travel, and to rising average temperatures from north to south, which drive increased evaporation. The northern parts of the HPA also receive more runoff from the Rocky Mountains.

Agriculture and Aquifer Depletion

High Plains Aquifer groundwater is pumped for municipal drinking supplies and for other domestic, commercial and industrial uses, but agriculture is far and away the largest consumer [Figure 2]. Irrigation in the High Plains was first implemented in the late 1800s, although extensive pumping of HPA groundwater for agriculture began only in the 1930s and 1940s in the aftermath of the Dust Bowl-era droughts that devastated farmland in parts of the region [Figure 3], particularly in the central and southern High Plains. Since then, and combined with more favorable climatic conditions in the latter half of the 20th century that brought rainfall at rates above long-term averages, groundwater has contributed substantially to the intensification of agriculture in the High Plains. Roughly 30 to 40 percent of High Plains cropland is now irrigated. It has also spurred

Figure 3: Farmland in Kansas devastated by wind erosion during the Dust Bowl.
Credit: USDA Natural Resources Conservation Service

shifts in cultivation, allowing crops with relatively high water demands, like corn, to be grown over wider areas than would otherwise be feasible.

Today, major crops grown in the High Plains include corn, cotton, sorghum, soybeans and wheat, among others. Groundwater use also supports livestock, which account for more than half of agricultural output by value over most of the region and consume much of the grain produced. That the High Plains, where annual precipitation is relatively low and surface water is generally scarce, have been so productive speaks to the perseverance of the farmers cultivating the land, as well as to the engineering feats that pump and transport fresh groundwater to keep crops growing through irrigation.

But the prolonged history of groundwater consumption has taken a toll. The total volume of water in the HPA prior to its development for groundwater pumping is thought to have been roughly 960 cubic miles (4,000 cubic kilometers), nearly as much as is in Lake Michigan. Estimates from the U.S. Geological Survey (USGS) and other researchers suggest the current volume is now 8 to 10 percent lower [Figure 4]. Although this amount of depletion — spread over more than half a century — may not sound especially alarming, the figure belies the tremendous variation seen in different parts of the region.

On average across the northern High Plains, which sits above roughly three-quarters of HPA groundwater,

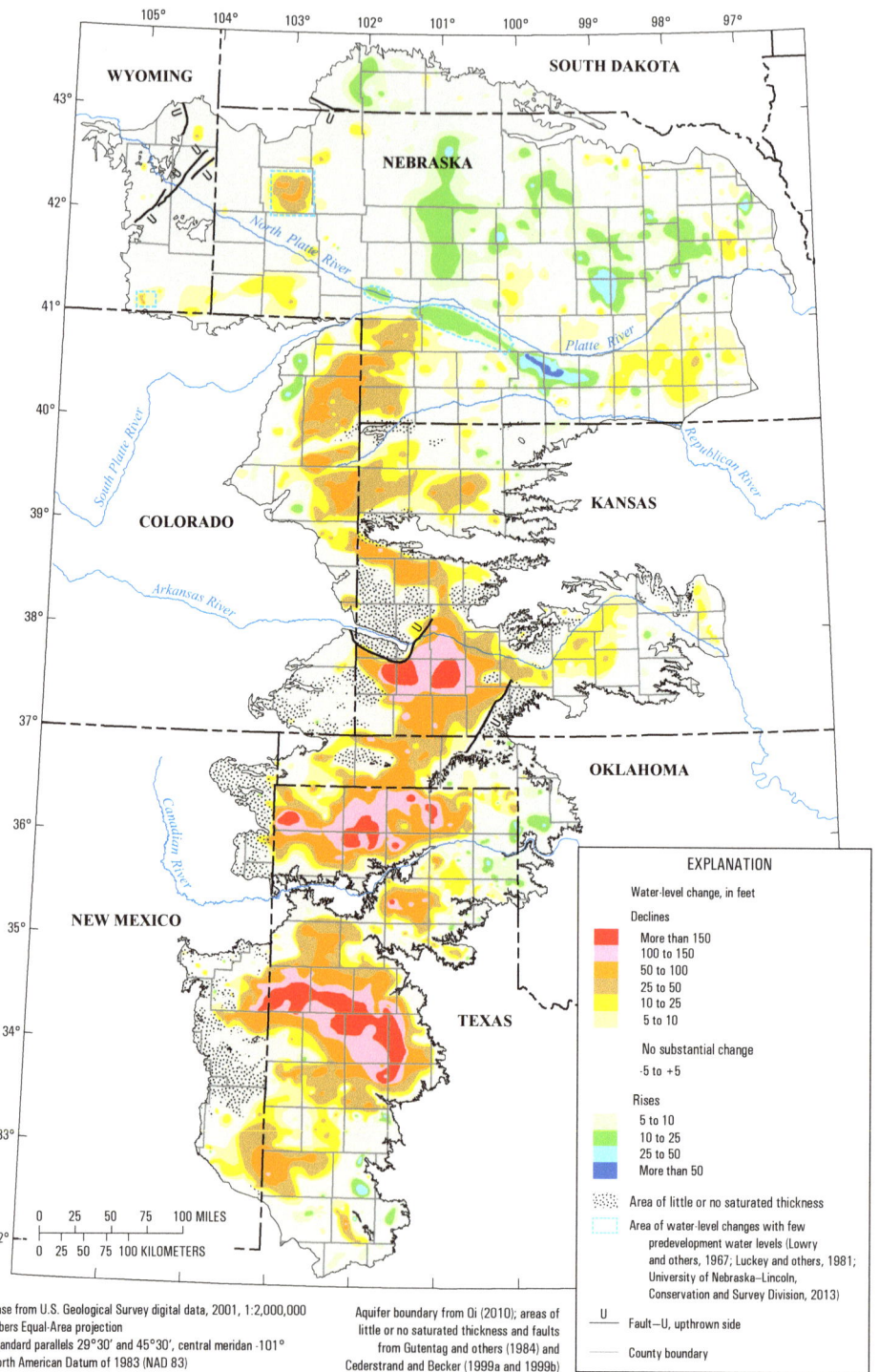

Figure 4a: Map of changes in the depth to the water table in the High Plains Aquifer (feet) from approximately 1950 to 2013.

Credit: U.S. Geological Survey Scientific Investigations Report 2014-5218, Virginia McGuire

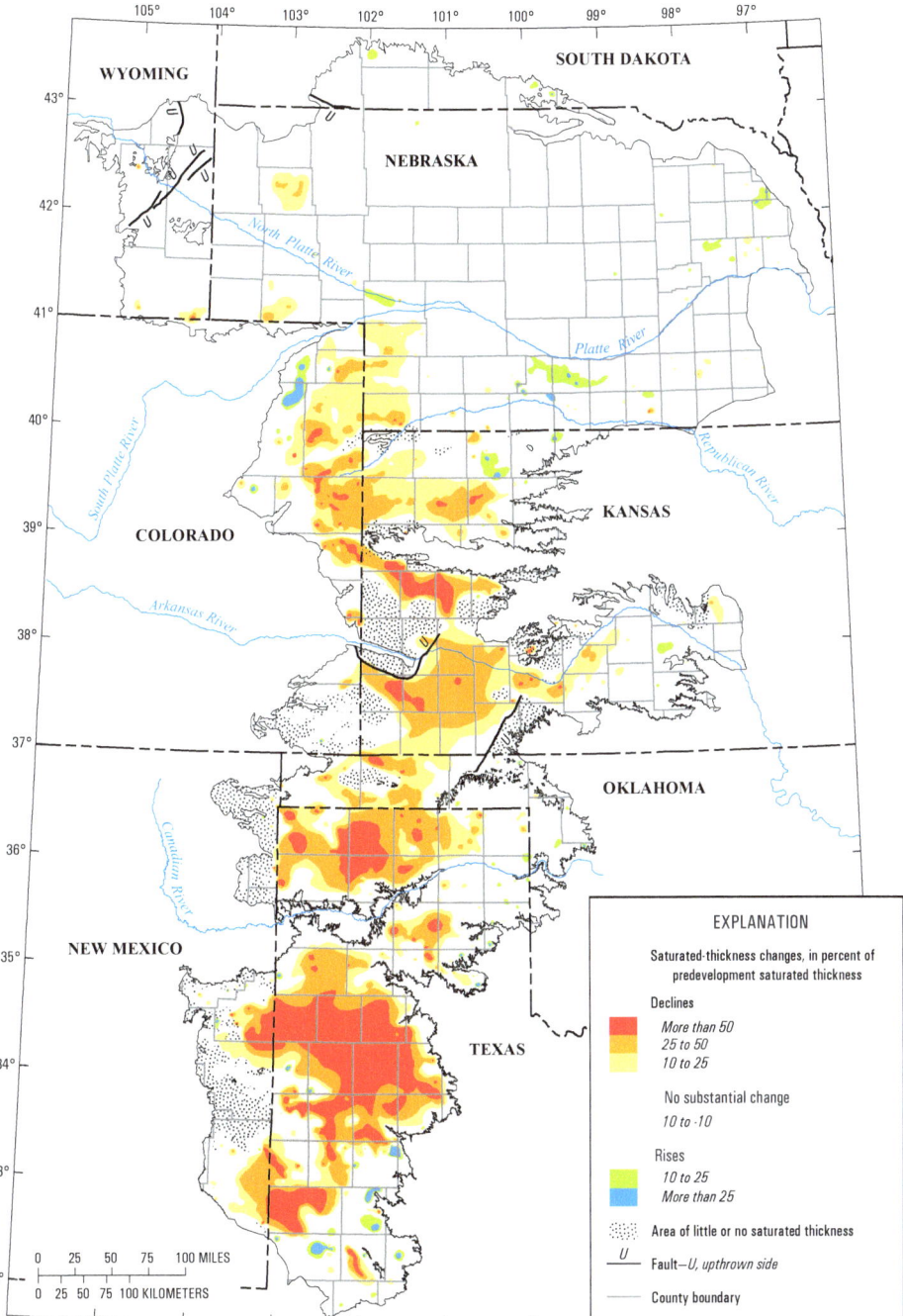

Figure 4b: Map of changes in groundwater thickness in the High Plains Aquifer from approximately 1950 to 2013.
Credit: U.S. Geological Survey Scientific Investigations Report 2014-5218, Virginia McGuire

> [In] portions of Kansas, Oklahoma, and Texas overlying the HPA ... water levels have dropped by about 13 to 41 feet (4 to 12.5 meters) on average.

there has been almost no depletion, and in parts of Nebraska, water levels have actually risen [Figure 4]. However, a different picture prevails across the portions of Kansas, Oklahoma and Texas overlying the HPA, where water levels have dropped by about 13 to 41 feet (4 to 12.5 meters) on average. The disparity is even greater at more local scales: The maximum drop in water level observed in Texas, for example, is nearly 263 feet (80 meters).

In the central and southern HP, where the initial volume of water in the aquifer was far smaller to start than in the northern HP and where discharge rates are five to 10 times higher than recharge rates, these drops in water level signify proportionally large depletions. Across large stretches of North Texas, as well as parts of western Kansas, groundwater use has depleted more than 50 percent of the aquifer's local saturated thickness, leaving relatively thin reserves of water at greater depth. The Kansas Geological Survey estimates, based on historical usage trends, that the aquifer will be effectively exhausted in large parts of western Kansas within 50 years, and some areas have already reached that point [Figure 5].

The lack of groundwater and the rising cost of pumping the diminishing supplies have forced farmers to adapt, fallowing or selling off land, switching to less water-intensive crops and/or dryland farming, or using more efficient irrigation technology. Such shifts, though necessary, usually come at a cost, whether through loss of agricultural land, cultivation of less profitable crops, decreased yields, or increased investment.

In addition to worries over how large-scale draw-downs in parts of the HPA will affect agriculture and livelihoods, there are other concerns associated with heavy groundwater use. Dropping water levels can lead to land subsidence as formerly saturated layers of the subsurface collapse under their own weight. Pumping and irrigation can at times introduce contaminants into groundwater, either via wells or by flushing salts in soil down into shallow, unconfined portions of aquifers. And even small reductions in groundwater levels can have significant effects, impacting flows in rivers, wetlands, and other surface waters that receive natural discharges from shallow groundwater sources. Reduced baseflows in the Platte and Republican rivers, which run predominantly through Nebraska (as well as through parts of Colorado, Kansas and Wyoming), due to small drops in the water table, have damaged ecosystems that are home to a number of endangered species and contributed to legal disputes over the use of the river flows.

In light of such ongoing issues of groundwater depletion and allocation, as well as projections for increased groundwater use in the future as demands on agriculture increase, temperatures rise, and droughts become more frequent and persistent, there is an urgency to reexamine existing governance frameworks and management plans concerning groundwater in the HPA — and likely elsewhere — to identify where and how they might be updated and improved.

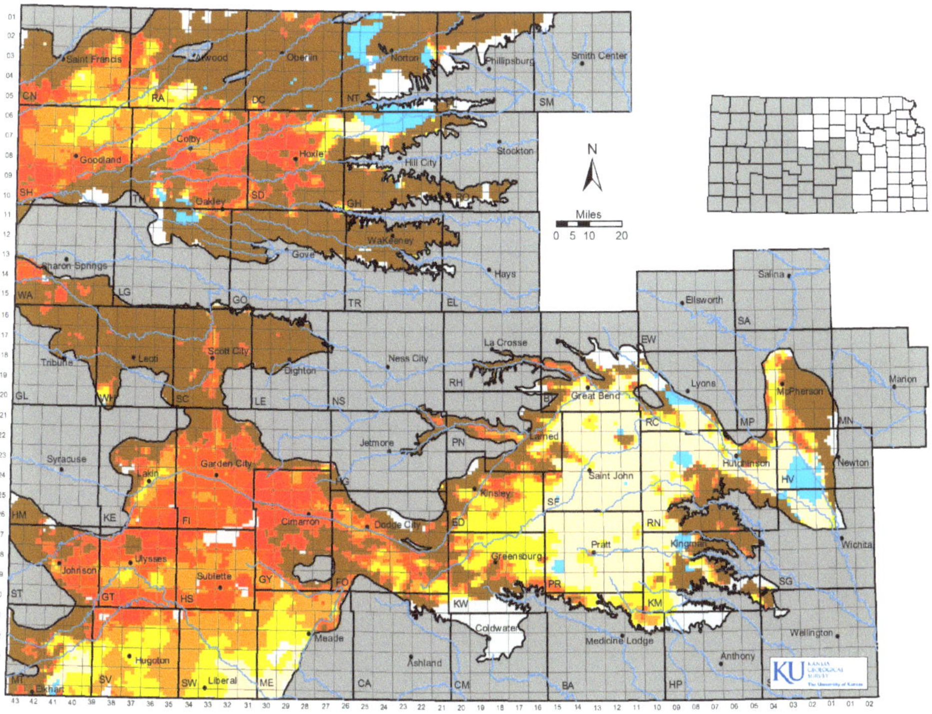

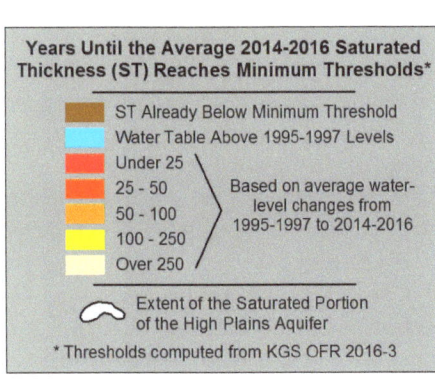

Figure 5: Map of estimated usable lifetime for the High Plains Aquifer in Kansas. (Based on groundwater trends from 1996–1998 to 2014–2016.)
Credit: Kansas Geological Survey Public Information Circular 18

Groundwater Governance and Management in the High Plains

Much like the details of groundwater availability and depletion, there is a mosaic of groundwater management and governance practices across the High Plains Aquifer region, with approaches varying considerably among and within states. This mosaic arose partly because groundwater resources and policies were developed at different times in different places, not all at once across each state or the entire region. The invisibility and lack of understanding of the HPA — particularly its finite size as well as how and over how long it is replenished — also contributed to a lack of broad-scale management early in its history. Communities of irrigators thus developed their own systems to manage local groundwater resources as they saw

> The states overlying the HPA have a diverse array of water laws that treat entitlements to groundwater use differently.

fit. By comparison, use of and impacts on surface waters are highly visible; hence, surface waters have been and continue to be regulated to a greater extent than groundwater.

Today, the states overlying the HPA have a diverse array of water laws that treat entitlements to groundwater use differently. In Texas and Oklahoma, for example, the right to pump groundwater is a property right attached to the land. Elsewhere, as in Kansas, water rights are owned, can be sold or passed on, and are prioritized based on when they were granted, but are not attached to the land. Each of the HPA states also have their own frameworks in place that parse the responsibilities of monitoring and regulating the aquifer to different extents between state agencies, like natural resources departments and geological surveys, and assortments of local to regional groups. In most states, groundwater management (or conservation) districts, natural resources districts, water conservation areas, and similar organizations allow local input into and some degree of autonomy — sometimes substantial — over the governance of groundwater.

In reality, the location-dependent details of groundwater availability, soil, climate, and other factors demand management tailored on a relatively small scale, so there is no ideal, one-size-fits-all setup for groundwater governance. And in the HPA region, it's clear that different approaches are more likely to succeed in different places depending on prevailing physical conditions, existing management frameworks, and cultural identities.

The most fundamental part of successful groundwater management is establishing clear management goals and desired outcomes. Although whether such goals are feasible, and what they might look like in practice, is often unclear. And reconciling competing goals is a major social, rather than technical, challenge. For many people, sustainability and the conservation of groundwater for future generations and/or to positively impact the

> There is no ideal, one-size-fits-all setup for groundwater governance.

environment are of utmost importance. For others, the managed use of water resources over set periods of time is the preferred goal, while others want to maximize the immediate return on investment.

Given the extremely slow rates of recharge of the aquifer throughout most of the central and southern High Plains — most of the water in those portions of the aquifer has been there for roughly 11,000 years — virtually any amount of pumping constitutes unsustainable mining of the resource (at least on human time scales) that will cause water levels to drop. This may seem tragic, but in these areas where groundwater is deep underground and doesn't directly impact surface waters or ecosystems, the question for managers — and for farmers, who are most directly affected by limitations on use and for whom irrigation equates to

The aquifer is close to the surface in north central Nebraska. Surface water flows in this area directly affect recharge rates. Conversely, groundwater levels often affect surface water flows.
Credit: Brenna Tobler

> The most fundamental part of successful groundwater management is establishing clear management goals and desired outcomes.

income and livelihoods — is whether there is actually any value in leaving the water underground. Although sustainability per se may be a tough sell in these areas, there are economic appeals for groundwater conservation that are perhaps more persuasive. For instance, if irrigated crops can be grown more efficiently in a crop-per-drop sense — through moderate reductions in water use or installation of new technology — a grower may be able to earn profit from farming irrigated land for longer.

Conservation may also help ensure that water will be available for successive generations to continue working family farms.

Elsewhere, such as across much of Nebraska and the northern HP, where the top of the aquifer is so close to the surface that it directly impacts surface water flows, sustainable use is a more common goal. High recharge rates also mean that sustainability is generally a more viable option, although managers still must balance the groundwater needs of irrigators to maintain crops.

Whatever the motivation — economic, environmental, or some combination thereof — effective management requires laying out quantifiable targets (e.g., for annual consumption) or desired future conditions (e.g., that a certain proportion of saturated thickness in a location will still remain in 50 years) along with clear plans for how to arrive at those conditions.

A Natural Resources Conservation Service (NRCS) soil scientist discusses nutrient management with a Saunders Co. farmer in Nebraska.
Credit: NRCS/Bob Nichols

Currently, most HPA states and/or local groundwater management agencies have stated goals of some sort, as well as roadmaps to achieve the goals based on available monitoring data and usage forecasts, both of which inform groundwater management practices. A key but often overlooked part of the success of such plans is flexibility. The timelines for management goals are typically laid out in decades. Over such periods, conditions — particularly with respect to weather and climate — can change dramatically, as can usage patterns, technology and our understanding of the aquifer. Thus, building in the capacity to adapt management goals, strategies and practices to changing conditions is likely a far better approach — and one that is practiced routinely in other industries and areas of natural resource management — than set-it-and-forget-it policies.

In addition to setting clear goals and adapting to change over time to achieve them, another critical component of effective groundwater management is the ability of the governance structure that's in place to actually carry out management plans. As mentioned, there are numerous different structures depending on where you look in the HPA; some have been more capable and more efficient than others at achieving goals such as reductions in use, maximizing the earning potential of groundwater use, and fostering communication among stakeholders. Occasional, critical assessment of current institutions can help ensure that administrative structures support groundwater management goals.

It's important to recognize, however, that as with any governing bodies, inertia develops the longer they exist and so there is built-in resistance, if not occasional hostility, to change — often both from the governing and the governed as they grow accustomed and attached to how things operate locally. Major adjustments to governance structures — like redrawing boundaries of established groundwater conservation districts or instituting new structures altogether — can also take substantial amounts of time to carry out, potentially disrupting resource management and counteracting intended benefits of making the adjustments. Thus, it's typically more expedient to look for ways to update existing governance frameworks and structures to effect desired changes in groundwater management.

Most existing governance frameworks in the HPA region are made up to some extent of institutions and stakeholders working on groundwater management at multiple levels, from the farm on up to the state government. These nested organizational structures take on many different appearances depending on location: In Oklahoma and parts of Texas, for example, irrigators are accountable directly to state-level oversight. Elsewhere in Texas and in most of Kansas overlying the HPA, groundwater management or conservation districts (often covering several counties, though their jurisdictions are not necessarily confined by county borders) oversee irrigators and are accountable to state agencies. Certain areas in Kansas are further designated as Intensive Groundwater Use Areas, Local Enhanced Management Areas, or Water Conservation Areas, each of which has additional implications for irrigators.

> Top-down water rules and regulations are likely to be unpopular; and the broader the imposing entity, the greater the objection

A lesson apparent from across much of the HPA is that nested governance frameworks are often necessary and generally effective, provided that responsibilities are appropriately allocated among institutions at different levels. A little consideration of how entwined water is in longstanding cultural identities — as well as in individual property rights — throughout much of the High Plains is enough to realize that top-down water rules and regulations are likely to be unpopular; and the broader the imposing entity, the greater the objection. Indeed, state regulatory efforts, and even local efforts, are generally met with opposition, sometimes fierce. Such opposition reduces cooperation among stakeholders and can be counterproductive in achieving groundwater management goals.

Beyond this, landowners naturally understand the nuances of the land they farm — what will grow where with how much water, which techniques may be most efficient, and so on — better than anyone. Thus, landowners, well drillers, and local management agencies are likely in the best position to facilitate community discussions about groundwater, to outline water management plans optimized for their area, and to oversee those plans once enacted.

Meanwhile, through legislation and education, states can help elevate groundwater issues in the public psyche; they can set broad targets for sustainability or efficiency improvements; and they can incentivize stakeholder efforts to achieve the goals. State-level entities, with their more substantial scientific, engineering and public relations resources, are generally better positioned to provide consistent statewide groundwater monitoring and forecasts of changing conditions to help guide decision-making, and to coordinate outreach and education efforts with and among local groups.

Enforcement efforts to ensure compliance with water laws and usage regulations are a responsibility that is sometimes best handled at the state level, or shared between the state and local agencies. While some groundwater management entities, such as Nebraska's Natural Resources Districts, monitor wells to check that irrigators are in compliance, local groups often do not have adequate resources to conduct thorough monitoring and/or to deal with instances of abuse or water rights disputes among irrigators. Furthermore, local policing can raise the potential for conflicts of interest and inconsistent treatment — whether willful or not — when those in charge of enforcement are more familiar with the constituents whom they are supposed to be both representing and monitoring. State agencies, on the other hand, typically have greater resources, as well as greater authority to act in cases of abuse or conflict; in theory, they should also be less subject to potential bias and should be able to provide more consistent enforcement.

> The rules by which all users are expected to abide... must be clearly defined and disseminated in collaboration with stakeholders.

Regardless of who handles enforcement, several factors are vital to the acceptance of such efforts by stakeholders, and to the success of these efforts in achieving broad compliance. First, the rules by which all users are expected to abide, as well as the range of violations and penalties for violations, must be clearly defined and disseminated in collaboration with stakeholders themselves. Second, if a rule exists, it must be enforced. And finally, enforcement and penalties must be applied fairly and consistently to all violators. Failure to meet these criteria can undermine the integrity of, and trust in, enforcement apparatuses, as well as in the broader groundwater governance efforts.

Another party that could play a role in nested governance frameworks is the federal government. Given the need for regionally and locally tailored management in the HPA, however, in addition to the high level of skepticism with which many irrigators and HPA stakeholders view federal oversight and intervention, such a role should be carefully considered and largely (or entirely) hands-off with respect to actual management of the resource. Several potentially useful areas for a federal presence in HPA governance include study of the aquifer and remaining resources, along with modeling and forecasting of future conditions, to complement state-level data collection and modeling; support for educational and outreach efforts related to groundwater resources; aligning agricultural policy with national goals for groundwater conservation; and offering incentives to states to pursue similar shared goals.

Measuring what You're Managing

Understanding the physical nature of aquifers and, even more importantly, accurately assessing groundwater volumes, usage and recharge rates, are foundational to successful groundwater management. Each of these requires collecting and analyzing data on both large and local scales. Although data don't create or present solutions to groundwater issues on their own, they are necessary in order for those involved in groundwater governance to make reasoned decisions about how to move forward. Further, in the absence of adequate data and information, there is increased uncertainty about the resource as well as more room for it to be negatively exploited.

Abundant data have been collected about groundwater in the HPA as a whole. The NASA-sponsored GRACE satellite mission collects large-scale

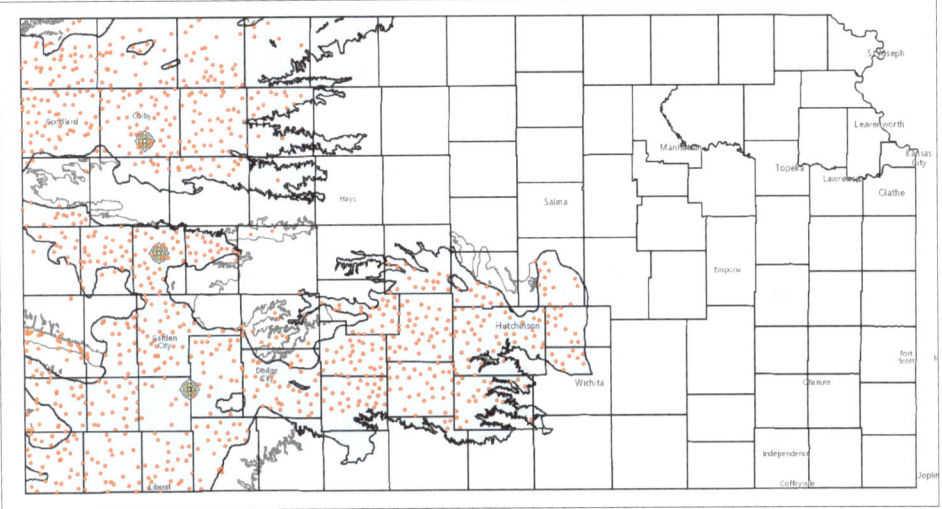

Figure 6: Map of Kansas groundwater monitoring wells (blue shading indicates High Plains Aquifer coverage; orange dots indicate monitoring wells).
Credit: Kansas Geological Survey

spatial data from space describing changes in the mass of groundwater stored beneath the High Plains (and elsewhere). And on the ground, the USGS and other federal agencies occasionally measure water levels from a selection of wells tapping the aquifer, as do state agencies and some local groups.

But as is often the case, where having some data is a good thing, having more is better — and this is certainly true in the HPA. Densely monitored water-level data are collected in some areas [Figure 6], but this is not the norm across the High Plains; at local scales, there are many knowledge gaps with respect to the condition and behavior of the HPA. Additionally, researchers who rely on occasional spot checks of water levels can only get snapshots of conditions in the aquifer at given times as opposed to continuous pictures throughout the year and from one year to the next. Furthermore, compared to water-level data, which offers a view of the aquifer's response to groundwater pumping, data on water usage — the other half of the aquifer use-response balance — is severely lacking across most of the region. Collecting these sorts of data on an ongoing basis can help illuminate seasonal cycles as well as short- and long-term trends in the aquifer.

Considering that groundwater management is ultimately intended to benefit irrigators and the public at large, however, collecting and analyzing more data, though necessary, is not enough. To be useful for policy makers, managers, farmers and the public, information about the aquifer must be communicated to these groups in forms they can easily access, digest and use. The need for landowners and other groundwater stakeholders to trust monitoring and management efforts is a major reason why data accessibility is important. Presenting information transparently, and being honest about the limits of current knowledge and both the facts and uncertainties of projections of future conditions and risk, will help build this trust.

> Outreach and public engagement... [are] imperative to the success of groundwater management goals.

The Kansas Geological Survey (KGS) offers a potential model for improved data collection and accessibility efforts. In 2007, the KGS began setting up a dense network of continuously monitored "index" wells tapping the HPA in the state [Figure 6]. Data from these wells (and from other wells monitored occasionally) is publicly available through the KGS website and has already helped illuminate seasonal and recent short-term trends in water levels near the wells. It has also been incorporated into economic impact studies and maps of estimated usable lifetime for the HPA across Kansas [Figure 5]; the maps suggest that the aquifer may be a viable source of groundwater for decades or centuries to come in some areas, while in other areas, the water remaining is already below the threshold needed to support agricultural pumping. Translating data into a more easily understood map in this way has made it easier for Kansas water managers to begin conversations with irrigators and the public about managing groundwater supplies.

Such efforts help demonstrate the practical value of data collection with respect to the HPA. But the logistics of expanding collection and communication of these data are tricky. Monitoring existing wells and installing new wells is expensive, time-consuming, and doesn't always align with budgetary priorities. In addition, considering that most wells are on private property, the level of trust that landowners have in the people and agencies tasked with monitoring can impact the ease with which wells can be accessed — even when an agency has a legal right to conduct monitoring on a property.

Building Trust through Outreach and Engagement

Making science and data from the HPA, along with useful data-based products, broadly accessible is a vital component of engaging with groundwater stakeholders, from legislators and local water management groups to individual irrigators and the public. And outreach and public engagement by the groups involved in groundwater data collection and governance, although time-consuming and often expensive, is imperative to the success of groundwater management goals. That's because monitoring, regulation, enforcement and other efforts aimed at achieving management goals cannot succeed without the support of stakeholders. And such support comes from the trust built through communication and dialogue between those governing and those governed.

Public engagement can and should take many forms as different people learn and respond best in different settings and with different approaches. Lecture- and classroom-style public presentations may be the most familiar means of community engagement and information sharing, but many other avenues exist. Online resources and tools, for example, let people process information at their own pace and convenience. Some groundwater management groups have developed incentivized certification programs and demonstration farms to introduce

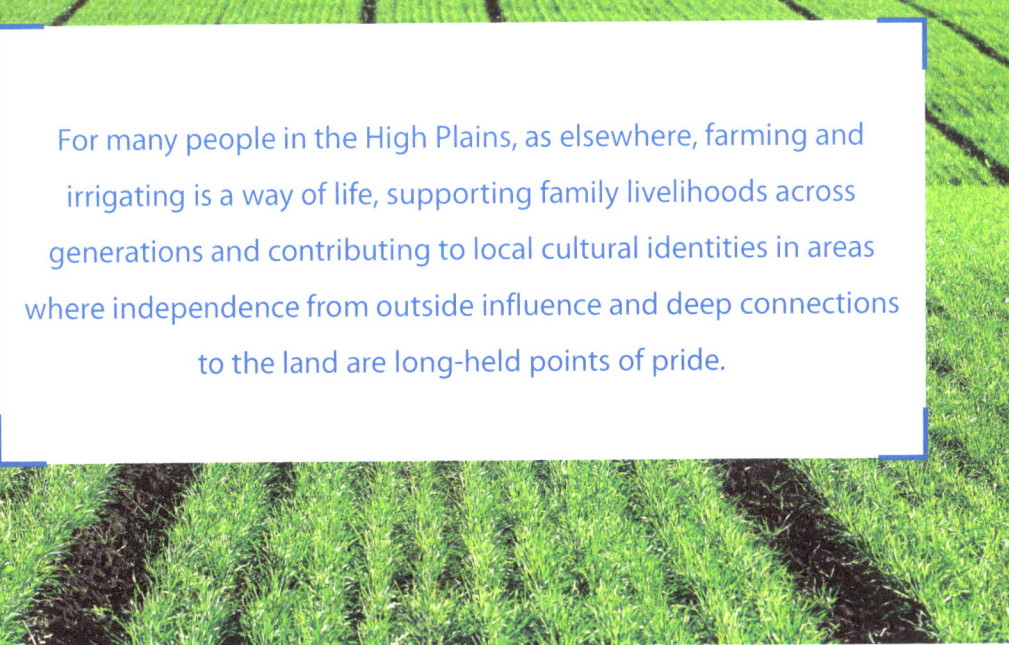

Credit: ©Shutterstock.com/AlexussK

irrigators to new practices and technologies designed to improve water-use efficiency. Including farmers and others in data collection — through citizen science programs or simply by involving irrigators in the process of checking well levels on their own land — is a means of engagement on

> Engagement should be approached as a conversation.

a more personalized level. Education efforts aimed at school-aged children and young adults (where possible, as school curricula are often subject to oversight by other groups) may also be effective means of communication, teaching students earlier on about how aquifers work and introducing them to current and potential future groundwater issues.

Another important element in effective public engagement with respect to groundwater is the way in which communication with an audience is handled. For many people in the High Plains, as elsewhere, farming and irrigating is a way of life, supporting family livelihoods across generations and contributing to local cultural identities in areas where independence from outside influence and deep connections to the land are long-held points of pride. Thus, outreach efforts in which communication only goes one way — with officials, for example, trying to educate farmers or prescribe new practices while not seeking input or feedback from the farmers themselves — face longer odds for success, particularly if those doing the talking are not from the community to which they're speaking.

Rather, engagement should be approached as a conversation: Those behind the outreach surely have messages they need to convey — about conservation measures, monitoring

> To have lasting positive impacts, public engagement efforts must be made often and on an ongoing basis.

efforts, future risk projections, etc. — but they should also solicit feedback from their audiences and then take this feedback seriously. After all, farmers, ranchers, well drillers and others, with their wealth of accumulated knowledge about how local lands, crops and climates behave, can often offer innovative strategies for groundwater use. Additionally, while officials should strive to offer honest portrayals of current and future conditions, information should be presented in ways that, as much as possible, eschew dramatized or threatening warnings about groundwater consumption and that refrain from assigning blame to specific groups or individuals. Overall, this approach promotes open dialogue and sharing of opinions and ideas, while discouraging potentially adversarial relationships between groundwater stakeholders and those tasked with managing resources.

Finally, to have lasting positive impacts, public engagement efforts must be made often and on an ongoing basis. If the people conducting outreach are frequently present and accessible in local communities — as opposed to stopping through once a year, say, to deliver an annual dose of education or perhaps unpleasant updates on the state of groundwater — they are more likely to earn the trust of the people they're trying to reach and to build faith in governance and management efforts.

To this end, local leaders and other trusted influential community members enlisted to do outreach and engagement, whether formally or informally, are often best positioned to make the most impact amid their peers. Such influencers may be found among irrigators who are early adopters of new technology and understanding; ranchers who rely on silage from farmers to feed their animals; university extension specialists; crop consultants; well drillers; and business owners and industry leaders who purchase from farmers or are otherwise involved or invested in local agriculture. Outreach efforts will never get through to nor resonate with all stakeholders. But, compared to organizations from outside a community, local leaders involved in their communities typically stand a better chance of identifying individuals and groups most likely to be receptive to engagement efforts, and thus maximizing the impact of these efforts.

A Need for Communication and Collaboration in Groundwater Management

Despite an emphasis during the Critical Issues Forum on the necessity for locally tailored groundwater management, as well as improved and increased data collection and outreach on local scales, forum participants also noted that the states and communities tapping the HPA are not all entirely unique from one another. Many places grow the same crops, have similar groundwater governance setups currently, and face similar present and future problems with respect to their groundwater resources. Furthermore, although the character of the aquifer varies from place to place, it is nonetheless a single system not confined by state or local borders; this means that the actions of one irrigator, community, or state can and often do impact others. Thus, an overall consensus that emerged from the forum was that there is abundant room — and need — for much more communication and collaboration among communities and local and state management agencies both within states and across state lines.

> Local leaders involved in their communities typically stand a better chance of identifying individuals and groups most likely to be receptive to engagement efforts, and thus maximizing the impact of these efforts.

Useful collaboration, for example, could come from sharing experiences with respect to successful (or unsuccessful) governance, management and public engagement approaches; from sharing collected data as well as tools developed to analyze data, forecast future aquifer conditions, and project risk; and potentially from groundwater compacts that provide opportunities for neighboring states or local agencies to work together — rather than at odds with one another — to achieve the management goals of each.

The road ahead for many people, communities, industries and ecosystems that rely on the High Plains Aquifer is unclear: How much water will be available, and for how long? And what's the best way to manage the resource for however long it is available? There is of course no one answer to these questions and no one-size-fits-all solution to groundwater issues throughout the HPA. However, communication, collaboration and information sharing offer ways for groundwater stakeholders to better understand the aquifer and learn about management practices that they perhaps had not previously considered. Put into practice, this greater awareness and knowledge could help prolong the availability and usefulness of groundwater in the High Plains.

©iStockphoto.com/DebiBishop

Speaker Biographies — Keynote Speakers

Sharon B. Megdal, Director, University of Arizona Water Resources Research Center
"Regional Aquifer Challenges in the West"

Sharon B. Megdal is Director of The University of Arizona Water Resources Research Center, an Extension and research unit in the College of Agriculture and Life Sciences. She also holds the titles: Professor and Specialist, Department of Soil, Water, and Environmental Science; C.W. & Modene Neely Endowed Professor; and Distinguished Outreach Professor. Her work focuses on water policy and management challenges and solutions, on which she writes and frequently speaks. Current projects include: comparative evaluation of water management, policy, and governance in growing, water-scarce regions; groundwater management and governance; groundwater recharge; and transboundary aquifer assessment. Sharon, who holds a Ph.D. in Economics from Princeton University, is active in several national water organizations and is an elected member of Central Arizona Project board, which is responsible for the rates, taxes, and policies of the largest surface water conveyance project in Arizona.

Merri Lisa Trigilio, Director/Producer, "Written on Water"
After-Dinner Keynote Talk

Written on Water's Producer and Director, Merri Lisa Trigilio, has an art degree in photography and film, and a doctorate in Geosciences from Penn State University. After fifteen years working as a geophysicist and later as a researcher in carbon sequestration, Merri Lisa found her way back to documentary storytelling. In 2012, she was a fellow at the Smithsonian Museum in Washington, DC, where she wrote and produced educational documentaries. She continues to explore science communication through the film medium, working as a freelance producer and director for educational institutions.

Jason Gurdak, Associate Professor, San Francisco State University
"Groundwater Policy in the Face of Climate Change"

Dr. Jason Gurdak is an Associate Professor of Hydrogeology in the Department of Earth & Climate Sciences at San Francisco State University. He is Coordinator of the UNESCO-International Hydrologic Program called Groundwater Resources Assessment under the Pressure of Humanity and Climate Change (GRAPHIC). GRAPHIC is a global-scale research, education, and outreach program that addresses climate change and sustainability of global groundwater resources. Prior to joining SFSU, he was a hydrologist for 11 years with the USGS. Dr. Gurdak has authored more than 50 publications in hydrology, including topics on the science and policy of climate change impacts and adaptation of groundwater resources.

Speaker Biographies

William M. Alley, Director of Science and Technology, National Ground Water Association

Dr. William M. Alley is Director of Science and Technology for the National Ground Water Association. He served as Chief, Office of Groundwater for the U.S. Geological Survey for almost two decades. Dr. Alley has published over 90 scientific publications and received numerous awards for his work, including the Meritorious Presidential Rank Award. He holds a B.S. in Geological Engineering from the Colorado School of Mines, an M.S. from Stanford University, and a Ph.D. from the Johns Hopkins University. He and his wife, Rosemarie, recently completed a general science book, "High and Dry," published in early 2017.

Jean Bahr, President, American Geosciences Institute

Jean Bahr has been on the faculty of the Department of Geoscience at the University of Wisconsin-Madison since 1987. She also participates in interdisciplinary graduate programs. Her research focuses on the interactions between physical and chemical processes controlling solute transport and transformation in groundwater systems. She is the current President of AGI and also an Editor of the American Geophysical Union's journal Water Resources Research. She was the 2003 Birdsall-Dreiss Distinguished Lecturer for Hydrogeology Division of the Geological Society of America (GSA) and served as GSA President in 2009-2010.

Ann Bleed, Former Director, Nebraska Department of Natural Resources

Ann Bleed, Ph.D., P.E. Emeritus, is retired, but currently is director on the Lower Platte South Natural Resources District. For most of her career Ann worked at the State of Nebraska Department of Natural Resources, first as the State Hydrologist, then as Deputy Director, and finally as Director of the Department. While at the Department she also served as a Nebraska representative on the negotiating teams that settled two interstate water allocation lawsuits over the North Platte and Republican Rivers before the U. S. Supreme Court, and helped develop the Platte River Recovery and Implementation Program.

Nick Brozovic, Director of Policy, Water for Food Global Institute at the University of Nebraska

Nick Brozovic is Director of Policy at the Water for Food Global Institute at the University of Nebraska. He works to ensure that the Institute's programs inform water management policies and decision makers. Brozovic has over 15 years of experience in water policy worldwide. A particular focus of his research is on evaluating policies and governance structures for agricultural water management, including water market design and implementation. He holds doctoral and master's degrees in agricultural and resource economics from the University of California-Berkeley, a master's degree in geology from the University of Southern California and a bachelor's degree in geology from Oxford University.

Speaker Biographies

Rex Buchanan, Director Emeritus, Kansas Geological Survey, University of Kansas

Rex Buchanan is the Director Emeritus of the Kansas Geological Survey, based at the University of Kansas. A native of Kansas, he is the co-author of Roadside Kansas: A Guide to its Geology and Landmarks (rev. edition, 2010) and editor of Kansas Geology: An Introduction to Landscapes, Rocks, Minerals, and Fossils (rev. edition, 2010), both published by the University Press of Kansas; and co-author of The Canyon Revisited: A Rephotography of the Grand Canyon, 1923-1991, published by the University of Utah Press (1994). He served as Secretary of the Association of American State Geologists and chaired the Kansas Task Force on Induced Seismicity. In 2008 he was named a fellow of the Geological Society of America and in 2016 received GSA's Public Service Award.

Jim Butler, Senior Scientist, Kansas Geological Survey, University of Kansas

Jim Butler is a Senior Scientist and Chief of the Geohydrology Section of the Kansas Geological Survey at the University of Kansas, where he has worked since 1986. He holds a B.S. in Geology from the College of William and Mary, and a M.S. and Ph.D. in Applied Hydrogeology from Stanford University. Jim was the 2007 Darcy Distinguished Lecturer of the National Ground Water Association and the 2009 recipient of the Pioneers in Groundwater Award of the Environmental and Water Resources Institute of the American Society of Civil Engineers.

Elizabeth Eide, Director, Board on Earth Sciences and Resources, The National Academies of Sciences, Engineering, and Medicine

Elizabeth Eide directs the Board on Earth Sciences and Resources and Water Science and Technology Board at the National Academies of Sciences, Engineering, and Medicine. The Boards oversee activities including energy and mineral resources; hazards; geotechnical engineering; geospatial and geographical science; and all issues related to water. Prior to joining the Academies in 2005, she was a research geologist for 12 years at the Norwegian Geological Survey. She is a Fulbright Scholarship recipient and was elected to the Royal Norwegian Society of Sciences and Letters. She completed a Ph.D. at Stanford University and B.A. at Franklin & Marshall College, both in geology.

James Eklund, Director, Colorado Water Conservation Board

James Eklund is the director of the Colorado Water Conservation Board (CWCB) and serves as Colorado's interstate representative on the Colorado River. As a lawyer and a government official, Eklund is already a disappointment to much of his family on the Western Slope. He is redeemed in their eyes, however, because he drinks whiskey and fights over water (but never at the same time). As the Director of the CWCB, Eklund leads the state's water policy, financing, and planning efforts. Eklund is a graduate of Stanford University and the University of Denver College of Law

Speaker Biographies

(neither of which, his father is quick to note, made him any better at cleaning ditches or irrigating pasture). The Upper Colorado River endangered fish he most identifies with is the Razorback Sucker because he thinks of himself as sharp but also somewhat gullible.

Wendy J. Harrison, Professor, Geology and Geological Engineering, Colorado School of Mines

Wendy J. Harrison is a tenured Professor of Geology and Geological Engineering at Colorado School of Mines. Her fields of scholarly expertise are in geochemistry and hydrology as well as geoscience education and she has published papers in topics that range from impact shock metamorphism in lunar materials, the formation of gas hydrates and their role in CO_2 sequestration, metals uptake by trees in mined lands, and mitigating respiratory quartz dust hazard. Dr. Harrison recently completed an appointment at the National Science Foundation as Division Director for Earth Sciences in the Geosciences Directorate. She currently serves as an academic advisor to the Petroleum Institute, Abu Dhabi and Nazarbayev University, Kazakhstan. Her work experience includes 8 years as a senior research geologist for Exxon Production Research Company in Houston, Texas.

John E. McCray, Professor and Head, Civil and Environmental Engineering, Colorado School of Mines

John McCray is Professor and Head of the Civil & Environmental Engineering Department at Colorado School of Mines, specializing in hydrology, water resources, and water quality. He is currently Mines PI of the NSF Engineering Research Center for urban water, ReNUWIt, the first ERC for water. He is a member of the U.S. EPA Science Advisory Board, a Fellow of the ASCE Environmental and Water Resources Institute, and was a Fulbright Fellow to Chile for water resources. He earned his Ph.D. in hydrology and water resources from the University of Arizona, and a BS in engineering from West Virginia University.

Kyle E. Murray, Hydrogeologist, Oklahoma Geological Survey, University of Oklahoma

Dr. Kyle E. Murray is a Hydrogeologist for the Oklahoma Geological Survey (OGS) at the University of Oklahoma (OU). His research covers a broad spectrum of topics in Oklahoma & the mid-Continent including water issues in the energy sector, regional water supply, contaminants of emerging concern (CEC), and wastewater reuse in the municipal and industrial sector. He is a member of the Oklahoma City Geological Society (OCGS), Geological Society of America (GSA), National Ground Water Association (NGWA), American Geophysical Union (AGU), and the International Association of Hydrogeologists (IAH) where he serves as an Associate Editor for Hydrogeology Journal.

Speaker Biographies

Susan Stover, Outreach Manager, Kansas Geological Survey, University of Kansas

Susan Stover, P.G., is Outreach Manager at the Kansas Geological Survey. She worked in water policy, water resource planning and environmental remediation for the State of Kansas for 20 years, before joining the Survey in 2014. Her experience includes working with stakeholders on programs and policies to conserve the High Plains Aquifer; organizing conferences on water and on teaching evolution; and hosting field trips for state legislators. She holds an M.S. in geology, University of Kansas, and a B.A. in geology, University of Nebraska. Stover is a Geological Society of America Fellow and vice-chair of GSA's Geology & Society Division.

Steven D. Walthour, General Manager, North Plains Groundwater Conservation District, Texas

Steve Walthour is the General Manager of the North Plains Groundwater Conservation District. He has 28 years experience in subsurface geology and groundwater management. Steve holds a Master's Degree from the University of Arkansas and is a licensed professional geoscientist in the State of Texas (License No. 1582).

David Wunsch, Director/State Geologist, Delaware Geological Survey

David R. Wunsch joined the Delaware Geological Survey as the new Director and State Geologist in November 2011. Dr. Wunsch came to DGS from the National Ground Water Association (NGWA), where he served as the Director of Science and Technology. Wunsch was the State Geologist of New Hampshire from 2000 to 2010, where he had statutory appointments to the New Hampshire Joint Board of Geology, and the NH Water Well Board, which oversee the licensing and adjudicatory proceedings for professional geologists and licensed well drillers, respectively. He is Licensed Professional Geologist in Kentucky, New Hampshire, and Delaware. Prior to his appointment as New Hampshire State Geologist, Dr. Wunsch was selected as the 1998-99 American Geological Institute Congressional Science Fellow.

About the American Geosciences Institute (AGI)

connecting earth, science, and people

The American Geosciences Institute represents and serves the geoscience community by providing collaborative leadership and information to connect Earth, science, and people.

AGI was founded in 1948, under a directive of the National Academy of Sciences, as a network of associations representing geoscientists with a diverse array of skills and knowledge of our planet. The Institute provides information services to geoscientists, serves as a voice of shared interests in our profession, plays a major role in strengthening geoscience education, and strives to increase public awareness of the vital role the geosciences play in society's use of resources, resilience to natural hazards, and the health of the environment.

AGI connects Earth, science, and people by serving as a unifying force for the geoscience community. With a network of 51 member societies, AGI represents more than a quarter-million geoscientists. No matter your individual discipline, AGI's essential programs and services will strengthen your connection to the geosciences.

EARTH Magazine: This monthly publication explores the science behind the headlines. EARTH magazine gives readers definitive coverage on topics from natural resources, energy, natural disasters and the environment to space exploration and paleontology and much more.

Education and Outreach: AGI Education offers products and services for K-12 educators, including NSF-funded curricula, high-definition videos, classroom activities, teacher professional development, and online resources.

GeoRef: GeoRef is a comprehensive, bibliographic database containing over 3.5 million references to geoscience journal articles, books, maps, conference papers, reports and theses.

Policy and Critical Issues: Geoscience Policy works with AGI member societies and policy makers to provide a focused voice for the shared interests of the geoscience profession in the federal policy process. Critical Issues provides a portal to comprehensive, impartial geoscience information for decision makers and holds frequent webinars to connect geoscientists to decision makers and the public.

Workforce: AGI produces the Directory of Geoscience Departments publication on human resources of the U.S. geosciences community. It collects data on the supply of and demand for geoscientists, and works with other organizations and government agencies to ensure that the health of the profession is understood.

Selected AGI Projects

Earth Science Week: Reaching over 50 million people a year, Earth Science Week promotes awareness of Earth science and appreciation of the geosciences' role in society. This international public awareness campaign, organized each October by AGI, provides informational resources, educational materials, and a variety of events and activities for students, teachers, and others. Program partners in government, industry, and the nonprofit sector unite to advance these efforts and continue the solid track record of success of this nearly two-decade-old initiative (www.earthsciweek.org).

Center for Geoscience & Society: The Center links geoscience information to diverse, non-specialist audiences, with a particular emphasis on communicating with decision makers at all levels and with educators in non-geoscience disciplines.

AGI's Geoscience Online Learning Initiative (GOLI): started in cooperation with the American Institute of Professional Geologists, provides asynchronous life-long and continuing education opportunities to the geoscience profession. GOLI provides live webinars, online courses via the OpenEdX platform, and continuous education credits for asynchronous learning.

AGI Foundation: The Foundation is the principal source of U.S. tax-deductible endowment and programmatic contributions to the American Geosciences Institute from industry, private foundations, and individual donors.

Critical Issues Program

The Critical Issues Program is a new program at the American Geosciences Institute. Its main purpose is to make geoscience information more discoverable to decision makers at all levels.

Critical Issues Website
www.americangeosciences.org/critical-issues
The Critical Issues website is a hub for decision-relevant, impartial geoscience information on many of society's most pressing issues. The Critical Issues website aggregates information from multiple geoscience organizations, making it easy for users to find trusted, comprehensive information from across the geosciences at one location.

The website's topic pages highlight resources from the geoscience community on climate, energy, hazards, mineral resources, and water, with easy-to-digest summaries, answers to common questions, portfolios of maps and tools, and links to more detailed documents about the issue.

Critical Issues Research Database
www.americangeosciences.org/critical-issues/research-database
The Critical Issues Research Database allows users to quickly search for topics, and link through to the documents on the websites of the organizations that produced the content.
- Contains more than 4,000 factsheets, reports, position statements, and case studies; expanding monthly
- Decision-relevant geoscience information, indexed for legislative staff and researchers
- Links users to the original source of the documents
- Searchable by location

Policy & Critical Issues Webinars
www.americangeosciences.org/policy-critical-issues/webinars
The Policy & Critical Issues programs host webinars on a variety of topics that bring geoscientists and decision makers together to discuss potential solutions to challenges at the interface of geoscience and society, including these past webinars:
- Water as One Resource: How interactions between groundwater and surface water impact water availability
- Desalination as a Source of Fresh Water
- Data as a National Asset for Decision Making
- Making Produced Water More Productive
- Assessing, Mitigating, and Communicating Flood Risk

Please follow us at:

 @AGI_GeoIssues

www.ingramcontent.com/pod-product-compliance
Lightning Source LLC
Chambersburg PA
CBHW040301220526
45473CB00002B/549